BRITISH
SIPUNCULANS

A NEW SERIES
Synopses of the British Fauna
No. 12

Edited by Doris M. Kermack

BRITISH SIPUNCULANS

Keys and Notes for the Identification of the Species

P. E. GIBBS

Marine Biological Association of the United Kingdom,
The Laboratory, Citadel Hill,
Plymouth PL1 2BP, England

1977

Published for
THE LINNEAN SOCIETY OF LONDON

by
ACADEMIC PRESS
LONDON, NEW YORK AND SAN FRANCISCO

ACADEMIC PRESS INC. (LONDON) LTD
24–28 Oval Road
London NW1 7DX

U.S. Edition published by
ACADEMIC PRESS INC.
111 Fifth Avenue
New York, New York 10003

Library of Congress Catalog Card Number: 77-77142
ISBN: 0-12-282050-9

Text set in 9/10 pt Monotype Times New Roman, printed by photolithography, and
bound in Great Britain by Henry Ling Ltd., Dorset Press, Dorchester, Dorset

Foreword

British Sipunculans is a small synopsis devoted to a phylum of bizarre marine invertebrates. It is the occurrence of small and varied groups of animals, like the sipunculans that makes the study of marine life so fascinating. Unfortunately these animals are often difficult to identify as their descriptions are usually to be found buried in large works, which may be unobtainable due to rarity or expense. It is the aim of the *Synopses of the British Fauna* to make this information easily available in the form of a pocket field or laboratory guide, which has spaces for the owner's notes and a waterproofed cover.

In the *Synopses* emphasis is placed upon the illustrations which are designed to show the whole animal, as well as of characters of particular diagnostic and systematic interest. Dr Peter Gibbs from the Marine Laboratory at Plymouth has illustrated his *Synopsis* with the clarity and skill of an artist and marine zoologist, who has devoted much time in the search, collection and description of these curious animals. This *Synopsis* will convey to naturalists, be they amateur or professional, his very real enthusiasm and interest in sipunculans.

In order to keep the production costs and therefore the retail price of the *Synopses* as low as possible, all authors donate their services and receive no fees. Not only the Linnean Society is grateful to them for this.

Further titles are now in press and specialist workers in other groups of animals have promised manuscripts in the future, so bookshops and book advertisements should be watched for the eventual appearance of further *Synopses of the British Fauna* (New Series).

Doris M. Kermack
Synopses Editor

A Synopsis of the British Sipunculans

P. E. GIBBS

*Marine Biological Association of the United Kingdom,
The Laboratory, Citadel Hill, Plymouth PL1 2PB, England*

CONTENTS

Introduction

The Phylum Sipuncula comprises a small but distinct group of vermiform animals, the body of which is unsegmented and composed of two main parts, a broadly cylindrical trunk and, anteriorly, a more slender, extensile introvert. Sipunculans are exclusively marine and do not penetrate estuaries to any extent. They occur in all seas and certain species are known to be widely distributed with records from worldwide areas. Their bathymetric range is wide; many species are to be found in intertidal situations and some extend to abyssal depths, the maximum recorded being about 7000 m. Sipunculans vary considerably in size; in small species the length of the trunk of mature specimens may be only a few millimetres whilst in the largest the trunk may measure over 0·5 m. Within the phylum, 16 genera and over 300 species are now recognized.

The British fauna, taken in this *Synopsis* to include those species recorded from the continental shelf down to a depth of 200 m, comprises six genera and 12 species, the majority of which can be found between the tide-marks. This fauna is essentially that of Scandinavian and European seas (see Théel, 1905; Cuénot, 1922; Fischer, 1925), although several species described from adjacent shelf areas, for example *Golfingia eremita* (Sars) and *Phascolion tuberculosum* Théel, have yet to be recorded from British waters. A further ten or more species occur in deeper water on the continental slope (for examples see Southern, 1913a).

1

General Structure

A sipunculan consists of a cylindrical, or pear-shaped **trunk** and a more slender **introvert** which is extended from, and withdrawn into, the anterior end of the trunk (Fig. 1A). When extended, the length of the introvert may be less than one-half up to about ten times the length of the trunk according to species. The anterior tip of the introvert, termed the **oral disk**, usually has a lobed **nuchal organ** on its dorsal side and carries **tentacles**, the shape, number and arrangement of which vary in different genera and species. Generally the tentacles are flattened or digitiform in shape and increase in number with growth. In *Golfingia* the tentacles are arranged in pairs around the **mouth** and usually the **tentacle crown** develops in two or three series with initially eight primary tentacles (Fig. 1B) followed by eight secondary and subsequently few to numerous tertiary tentacles; in the adult the tentacle crown can consist of tentacle pairs arranged in a single-tiered circle, as in *G. elongata* (Fig. 1C), or it can be more complex with paired radial/longitudinal rows of tentacles which may number 100 or more, as in *G. vulgaris* (Fig. 1D). In *Phascolosoma* and *Aspidosiphon* the tentacles are arranged in a horseshoe-shaped crescent enclosing the nuchal organ (Fig. 1E) whilst in *Sipunculus* the tentacular crown takes the form of a flat tentacular fold encircling the mouth (Fig. 1F). Behind the disk a zone of chitinous, posteriorly-directed **hooks** may be present, arranged either in well marked rings or scattered in an irregular fashion; in golfingiids the hooks are simple and spine-like (Fig. 1G), whilst in others they are flat, curved structures. The **anus** is situated on the mid-dorsal line towards the anterior end of the trunk (except in *Onchnesoma* where it is removed to the introvert) and, at about the same level, one or two **nephridiopores** are present, usually ventrolateral in position. The **skin** contains groups of glandular cells, forming the **epidermal organs**; these are often raised to form papillae which vary in size and shape and in some species they form useful diagnostic features. In some genera the skin at one or both ends of the trunk is heavily chitinized forming **shields** e.g. *Aspidosiphon*. The **body wall** is composed of two main **muscle layers**, an outer circular and an inner longitudinal; these layers may be continuous sheets [Fig. 1A(1)] or collected into bands [Fig. 1A(2) (3)]. In some genera, e.g. *Sipunculus*, the body wall contains longitudinal canals or sacs which communicate at intervals with the coelom.

Fig. 1. A. External characters of a sipunculan: circular and longitudinal muscle layers of body wall may be both in continuous sheets (1) as in *Golfingia* spp., or both in bands (3) as in *Sipunculus*, or mixed, with circular continuous, longitudinal in bands (2) as in *Phascolosoma*; B–F. Diagrams illustrating the structure of the oral disk (m.—mouth; n.o.—nuchal organ): B. Simple form with primary tentacles only (juvenile *Golfingia* type); C. *G. elongata* type with tentacles in a single-tiered circle around mouth; D. *G. vulgaris* type with tentacles in radial/longitudinal rows around mouth; E. *Phascolosoma–Aspidosiphon* type with tentacles enclosing nuchal organ; F. *Sipunculus* type with large tentacular fold around mouth; G. Top and side views of simple, spine-like hook found in *Golfingia* and other genera; H. Main internal structures of a generalized sipunculan with introvert withdrawn, dissected from dorsal side; J. Trochophore larva of *G. vulgaris* (after Gerould, 1906).

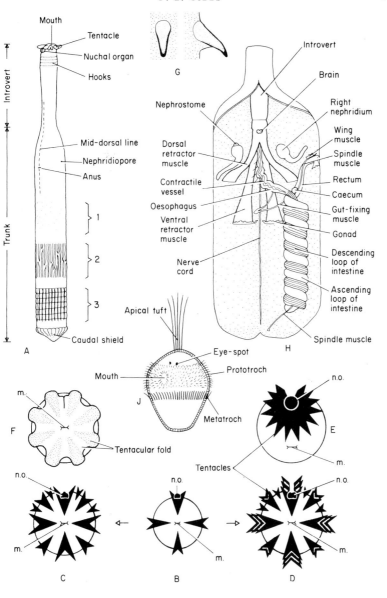

A — Mouth, Tentacle, Nuchal organ, Hooks, Introvert, Trunk, Mid-dorsal line, Nephridiopore, Anus, 1, 2, 3, Caudal shield

G

H — Introvert, Brain, Right nephridium, Wing muscle, Spindle muscle, Rectum, Caecum, Gut-fixing muscle, Gonad, Descending loop of intestine, Ascending loop of intestine, Spindle muscle, Nephrostome, Dorsal retractor muscle, Contractile vessel, Oesophagus, Ventral retractor muscle, Nerve cord

J — Apical tuft, Eye-spot, Mouth, Prototroch, Metatroch

F — m., Tentacular fold

Tentacles

E — n.o., m.

C — n.o., m.

B — n.o., m.

D — n.o., m.

Internally, the main muscles are the **retractors** of the introvert; these are longitudinal bands which cross the coelom from the base of the oral disk to various levels of the trunk wall (Fig. 1H). The number and arrangement of the retractor muscles is of generic and subgeneric importance; there may be four retractors arranged in dorsal and ventral pairs, as in *Golfingia* (*Golfingia*) spp.; or two, one dorsal and one ventral (=fused pairs), as in *Phascolion* spp.; or two in a ventral pair, as in *Golfingia* (*Phascoloides*) spp.; or only one (fused ventral pair) as in *Aspidosiphon* spp. Within a species, the retractor muscle system often shows considerable variation, particularly in the degree of fusion of the ventral muscles, and marked abnormalities, such as the non-development of one or two muscles, are frequently encountered.

The **gut** consists of three, ill-defined sections, an **oesophagus**, a long **intestine** which in many species is coiled in a double spiral, and a **rectum** generally bearing a **caecum**. The double spiral of the intestine consists of a "descending" loop and an "ascending" loop and is supported by an axial strand, the **spindle muscle**, attached near the anus and sometimes also to the posterior end of the trunk. The gut is usually anchored by slender **fixing muscles**, fastening the oesophagus and gut spiral to the trunk wall, and the rectum is fastened on either side of the anus by broad attachments known as **wing muscles**. The tentacles operate by means of a compensatory system in the form of a **contractile vessel** which is closely fastened to the oesophagus; in most species this vessel is a simple tube but in *Sipunculus* it is double and in *Golfingia* (*Thysanocardia*) spp., it has numerous villi. The **nephridia** are two sac-like structures in the anterior trunk region, often attached to the trunk wall by fine strands; each opens to the coelom by an anterior **nephrostome**. In some genera (e.g. *Phascolion*) only one **nephridium** (usually the right) is developed. The **nerve cord** is mid-ventral in position, running anteriorly to the **brain**, a bilobed mass situated dorsally behind the oral disk. The **gonads** appear as fine strands at the base of the ventral retractors; gametes are released to the coelom at an early stage of development and prior to the spawning season, often completely fill the coelom.

More detailed accounts of the structures and anatomy of sipunculans will be found in Hyman (1959) and Tétry (1959).

General Biology

Sipunculans are sedentary animals commonly found around the British coast
~~i~~ muddy sands and gravels on the lower shore and in the sublittoral zone. Large
~~i~~ecies, such as *Golfingia elongata* and *G. vulgaris*, construct vertical burrows
~~-~~hich may penetrate to a depth of 30–50 cm; smaller species inhabit the surface
~~i~~yers, often utilizing the empty shells of gastropods and scaphopods or other
~~i~~milar shelters. This habit is adopted by *Phascolion strombi*, probably the com-
~~i~~onest and most widespread of European Sipuncula. On rocky shores the small
~~i~~. *minuta* may be abundant in sedimented crevices or similar niches. In the
~~-~~opics many species bore into corals and coralline rock. Most sipunculans
~~i~~ppear to be predominantly deposit feeders although a seemingly parasitic
~~i~~ehaviour has been observed in *G. procera* (Thorson, 1957). They figure in the
~~i~~iets of many fish and some species are used as fishing bait (e.g. *Sipunculus nudus*);
~~i~~ the tropical Pacific region several of the large species are eaten by man.

With one exception, in all sipunculans investigated to date, the sexes are
~~i~~eparate but external dimorphism is lacking; in most species the sexes are present
~~i~~ about equal numbers. Generally, spawning occurs over a limited period of the
~~i~~ear, the gametes being liberated from the coelom via the nephridia and fertilized
~~i~~xternally. The resulting embryo develops to a **trochophore larva** with proto-
~~i~~ochal and metatrochal ciliated bands (Fig. 1J) and either leads a brief planktonic
~~i~~xistence lasting several days before metamorphosis or undergoes further
~~i~~evelopment to a **pelagosphaera stage** lasting a month or more (see Rice,
~~i~~975). The exception noted above is *Golfingia minuta* which is perhaps unique
~~i~~mongst the Sipuncula in being an hermaphrodite; in this species the larva is
~~i~~etained in the burrow of the adult throughout its direct development (see
~~i~~kesson, 1958; Gibbs, 1975).

As with many other sedentary groups, sipunculans attract a variety of asso-
iates and parasites. A search of the burrow of *Golfingia elongata* or *G. vulgaris*
~~i~~ay reveal the small bivalves *Mysella bidentata* (Montagu) and *Epilepton
clarkiae* (Clark) or the polychaete *Harmothoe lunulata* (delle Chiaje). Another
~~i~~olychaete, *Syllis cornuta* Rathke, cohabits shells occupied by *Phascolion strombi*
~~i~~nd *Aspidosiphon muelleri*. Epizoitic entoprocts of the genus *Loxosomella* are
~~i~~ommonly attached to *G. vulgaris* and *P. strombi*. External parasites are less
~~i~~requently encountered and include the copepods *Myzomolgus* and *Catania* on
~~i~~ipunculus nudus* and the gastropod *Menestho diaphana* (Jeffreys) (=*Odostomia
~~i~~erezi* Dautzenberg and Fischer) on *Phascolion strombi* (see Kristensen, 1970).
~~i~~nternal parasites include various protozoans, trematodes and a copepod,
~~i~~kessonia occulta* Bresciani and Lützen, the latter occurring in *Golfingia minuta*.

Collection and Preservation

On the lower shore the deeper burrowing species (e.g. *Golfingia elongata*, *G. vulgaris*) inhabiting muddy deposits can be collected by digging with a fork to a depth of 15–30 cm; offshore, a heavy dredge of the "anchor" type (Forster, 1953) is usually required for such species but for shallow burrowing species (e.g. *G. procera*, *Phascolion strombi*) a "naturalist" dredge or a grab is normally adequate. Crevice-dwelling species (e.g. *Golfingia minuta*, *G. rimicola*) are readily obtained by carefully prising open sedimented fissures with a crowbar. When possible species living in shells should be removed from the shelter before preservation in order to facilitate rapid penetration of the fixative; specimens are often firmly fixed inside shells and thus it is better to break the shells using a hammer or a vice rather than attempt to remove specimens by pulling.

The external characters of the introvert are important for identification and most easily determined when the introvert is fully extended. In some species this state of preservation can be rather difficult to obtain since different species do not react in the same manner when given a similar treatment. Best results are always obtained with specimens that have been maintained in cool conditions since the time of collection. However, in *Golfingia* spp. for example, the most successful method seems to be placing the specimens in a mixture of sea-water and dilute alcohol (approx. 10%) in equal parts; in this mixture, specimens generally slowly relax to a point where the introvert is not withdrawn when stimulated (after 10–20 min) and can then be transferred to fixative. If results are poor with this method, other standard methods of narcotization can be tried, such as the addition of menthol crystals or immersion in a solution of either magnesium chloride (7% made up in tap-water or distilled water) or propylene phenoxetol (1%). For identification purposes, alcohol (70%) is a convenient fixative since it rapidly kills and hardens specimens. Alternatively, neutral formol (5%) can be used; this fixative causes less coagulation of the coelomic contents and thus is better than alcohol for studies of the internal details.

The identification of sipunculans requires an examination of the internal structures, particularly of the number and arrangement of the retractor muscles. This is a simple operation involving a longitudinal incision of the dorsal body wall along the length of the trunk to one side of the anus. The body wall is then pinned out and coelomic contents (gametes may fill the coelom of mature specimens) washed out. Careful manipulation of the gut coils will enable the observer to determine all important taxonomic details.

In life, the body colouration of sipunculans varies from pearly white (e.g. *Sipunculus nudus*), through shades of light to dark, greyish or reddish, brown (e.g. most *Golfingia* spp.) to black (e.g. some *Aspidosiphon muelleri*). These hues are generally retained after preservation but, for the most part, are of little use in distinguishing the species.

Classification

PHYLUM SIPUNCULA
Family Sipunculidae
  Sipunculus nudus Linnaeus, 1766
Family Golfingiidae
  Golfingia (Golfingia) elongata (Keferstein, 1862)
  Golfingia (Golfingia) margaritacea (Sars, 1851)
  Golfingia (Golfingia) vulgaris (de Blainville, 1827)
  Golfingia (Phascoloides) minuta (Keferstein, 1862)
  Golfingia (Phascoloides) rimicola Gibbs, 1973
  Golfingia (Thysanocardia) procera (Möbius, 1875)
  Phascolion strombi (Montagu, 1804)
  Onchnesoma squamatum (Koren and Danielssen, 1876)
  Onchnesoma steenstrupi Koren and Danielssen, 1876
Family Aspidosiphonidae
  Aspidosiphon muelleri Diesing, 1851
Family Phascolosomatidae
  Phascolosoma granulatum Leuckart, 1828

Systematic Part

The classification of the Sipuncula has a confused history. Formerly, these animals were placed in the Phylum Gephyrea, a group of uncertain affinities comprising the Classes Sipunculoidea, Echiuroidea and Priapuloidea. However, the concept of the Gephyrea is no longer held to be tenable, and whilst sipunculans show clear embryological affinities with the Annelida, it is now generally agreed that, because of their total lack of segmentation, they are best regarded as a distinct and separate phylum. The name Sipuncula, proposed by Stephen (1964), appears to have been accepted generally; the common name becomes 'sipunculan" instead of "sipunculid", the latter being reserved for a species within the Family Sipunculidae. Whilst attempts have been made to group the various genera, no formal system of higher categories was proposed until 1972 when Stephen and Edmonds distinguished the four families listed above. All four families are represented in the British fauna.

Considerable morphological variation is shown by some sipunculans and as a consequence many different names have been applied to certain species. Full lists of the synonyms of the species described below will be found in Stephen and Edmonds (1972). It should be noted that the priority of the generic name *Golfingia* Lankester was not established until 1950 when Fisher reviewed the genus *Phascolosoma* Leuckart. Prior to this date, the name *Phascolosoma* had been misapplied to *Golfingia* spp. and other names used for *Phascolosoma* spp. Recently, Cutler and Murina (1977) have reviewed the genus *Golfingia*; they have elected to retain the subgeneric name *Phascoloides* Fisher rather than replace it with the name *Nephasoma* Pergament, which has priority (Murina, 1975), since unnecessary confusion would result in such a major nomenclatural change. This view has been adopted in the present work.

The size of individual species cannot be stated precisely since the parameters of the body alter according to the state of contraction and degree of introvert protrusion. The body lengths given in the descriptions below are approximations of the total length of specimens preserved with the introvert fully extended.

All the keys in this synopsis are applicable to the British representatives only.

Key to Families of British Sipuncula

1. Trunk with a hard chitinous shield at either end .   .   Aspidosiphonida
                                           (*Aspidosipho*

     Trunk without shields   . . . .

2. Longitudinal muscle layer continuous   . . . . .   Golfingiida

     Longitudinal muscle layer collected into bands . . . . . .

3. Oral disk with large tentacular lobe surrounding mouth; introvert
     without hooks . . . .   Sipunculida
                                            (*Sipunculus*

     Oral disk with tentacles arranged in a crescent enclosing nuchal organ;
     introvert with hooks . . . .   Phascolosomatida
                                           (*Phascolosoma*

Oral disk with tentacles surrounding mouth; tentacles of variable form, few
o numerous, sometimes rudimentary or even absent. Coelomic canals or sacs
ot present in body wall. Longitudinal muscle layer continuous, not collected
nto bands. Trunk without shields. One or two nephridia. The bulk of the
British fauna belongs to this family represented by three genera and nine species.

. Trunk with ∩ or ∧ -shaped "holdfast" papillae (Fig. 8D); two retractors
   —one dorsal, one ventral; usually in gastropod shell
                                                 Phascolion strombi (p. 22)
   Trunk without "holdfast" papillae, one to four retractors **2**

. One retractor; one nephridium; anus on introvert (Fig. 9D) **3** (*Onchnesoma*)
   Two to four retractors; two nephridia; anus on anterior trunk (Fig. 2D)
                                                        4 (*Golfingia*)

. Trunk covered with prominent, backwardly-directed protrusions (Fig. 9C);
   about eight tentacles . . . . . *Onchnesoma squamatum* (p. 24)
   Trunk with flat to spherical "scales" (Fig. 10A); no tentacles
                                          Onchnesoma steenstrupi (p. 26)

. Two retractors **5**
   Four retractors (sometimes three or two) (subgenus *Golfingia*) . . . **7**

. Contractile vessel with numerous villi (Fig. 7E) (subgenus *Thysanocardia*)
                                                 Golfingia procera (p. 20)
   Contractile vessel without villi (subgenus *Phascoloides*) **6**

. Nephridiopores anterior to anus (Fig. 6A); hooks arranged regularly in
   rings (Fig. 6C); up to 20 tentacles . . . *Golfingia rimicola* (p. 18)
   Nephridiopores posterior to anus (Fig. 5A); hooks (when present) not
   arranged regularly (Fig. 5C); two tentacles . *Golfingia minuta* (p. 16)

. Introvert without hooks . . . . . . *Golfingia margaritacea* (p. 12)
   Introvert with hooks **8**

. Hooks arranged regularly in rings (Fig. 2C); tentacles (up to 36)
   arranged in single-tiered circle (Fig. 2B) . *Golfingia elongata* (p. 10)
   Hooks not arranged regularly (Fig. 4C); tentacles (up to 150) arranged
   in radial/longitudinal rows (Fig. 4B) . . *Golfingia vulgaris* (p. 14)

Golfingia (*Golfingia*) *elongata* (Keferstein)

(Fig. 2A–D)

Phascolosoma elongatum Keferstein, 1862
Golfingia (*Golfingia*) *elongata*: Stephen and Edmonds, 1972: 90

A large species, up to about 140–150 mm in length, with a slender, cylindrica trunk (Fig. 2A). Oral disk carries a single-tiered circle of paired tentacles sur rounding mouth (Fig. 2B); number of tentacles in adults (>30 mm long) varie between 20 and 36, the majority having 24–28, but juveniles may have as few a eight. Simple, spine-like hooks are arranged in well separated rings on anterio introvert; juveniles have up to 20 rings (Fig. 2C) but adults may have only thre or four as a result of wear. Skin of introvert and trunk fairly smooth, withou prominent papillae. Nephridiopores are ventrolateral and just anterior of anu on anterior trunk. Longitudinal muscle layer of body wall is continuous, nc collected into bands.

Internally, four retractor muscles are normally present, the ventral pair in serted in middle third of trunk and the dorsal pair at about level of anus (Fig 2D); one or both of the dorsal retractors may be missing in aberrant specimen (see Notes below). Intestine is tightly coiled in a double spiral supported by spindle muscle attached anteriorly near anus but not posteriorly. Rectal caecur usually conspicuous. Gut-fixing muscles fasten oesophagus on left and rectur on right to body wall in the region between roots of dorsal and ventral retractors frequently one (or both) of these muscles is missing. Contractile vessel simple Two nephridia; these hang freely in coelom.

Inhabits muddy sand/gravel from lower shore to about 170 m depth. Juvenile are occasionally found in rock crevices on lower shore, together with *Golfingia rimicola* (p. 18) and *G. minuta* (p. 16).

A common and widespread species in British waters, found from the Skagerra to the eastern Mediterranean. Also reported from Cuba.

Type locality: St. Vaast la Hougue, France.

Notes. Aberrant individuals of *G. elongata* with only three or two retracto muscles (one or both of the dorsal pair not developed) are frequently encoun tered. In the Plymouth area, such aberrant specimens comprise up to 15% of some populations (Gibbs, 1973). Those individuals with only two retractor may be confused with *G. rimicola* but generally *G. elongata* can be distinguished b the greater number of tentacles (22 or more) compared with *G. rimicola* (18 o less) (p. 18).

G. elongata spawns in July and August (Plymouth). Its burrows are occasionally inhabited by the polychaete *Harmothoe lunulata* and the bivalves *Epileptor clarkiae* and *Mysella bidentata*.

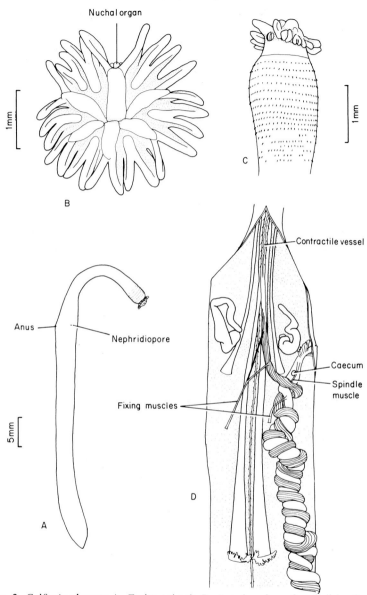

FIG. 2. *Golfingia elongata* A. Entire animal; B. Anterior view of oral disk of adult specimen with 26 tentacles; C. Dorso-lateral view of anterior introvert of young specimen (22 mm in length); D. Internal structures of anterior trunk region.

Golfingia (Golfingia) margaritacea (Sars)

(Fig. 3A–D)

Sipunculus margaritaceus Sars, 1851
Golfingia (Golfingia) margaritacea: Stephen and Edmonds, 1972: 94

A highly variable species in size, shape and morphology; littoral specimens are small, not exceeding 20 mm in length (Fig. 3A) whilst those from the sublittoral may attain lengths of 100–300 mm. Trunk is cylindrical, tapering posteriorly, sometimes markedly so forming a long, appendix-like "tail" (see Wesenberg-Lund, 1930). Tentacular crown becomes more complex with increasing size. In small specimens, a single circle of tentacles surrounds mouth (Fig. 3B) as in the littoral form of British coasts which typically has eight to 16 tentacles but no more than 20; in larger sublittoral forms (particularly from more northern latitudes), the number of tentacles is considerably greater with additional tentacles developing in radial/longitudinal rows around the margin of the oral disk in the largest specimens over 100 may be present (Fig. 3D). No hooks on introvert. Skin of introvert and trunk finely wrinkled, with minute papillae; in large specimens skin often has a lustrous, "mother-of-pearl" appearance. Nephridial pores are ventrolateral, just anterior of anus on anterior trunk. Longitudinal muscle layer of body wall is continuous, not collected into bands.

Internally, four retractor muscles are present, the ventral pair inserted in middle third of trunk, the dorsal pair at about, or posterior to, level of anus (Fig. 3C). Intestine is tightly coiled in a double spiral supported by a spindle muscle attached anteriorly near anus, but not posteriorly. Rectal caecum usually well developed. Gut-fixing muscles fasten oesophagus and anterior coils of gut spiral to body wall in the region between roots of dorsal and ventral retractors. Contractile vessel simple, not readily seen in small specimens. Two nephridia, these hang freely in coelom.

Inhabits mud and sand/gravel from lower shore (crevices) to about 4600 m depth.

In British waters this species is uncommon. Records are confined to the littoral zones of south-west England and western Scotland, plus a few scattered localities in the central and northern North Sea (Gibbs, 1974). It has a worldwide distribution, mainly in temperate and polar regions.

Type locality: Finmark (Hammerfest, Komagsfjord), northern Norway.

Notes. G. margaritacea has a long list of synonyms and a considerable number of subspecies are recognized although the morphological constancy and nomenclatural value of these forms is questionable (see Stephen and Edmonds, 1972; Cutler, 1973). Significantly, large, presumably old, specimens show the greatest morphological variation, particularly in the structure and appearance of the skin.

Around south-west England, this species matures at a small size (<10 mm in length) and probably spawns in October or November.

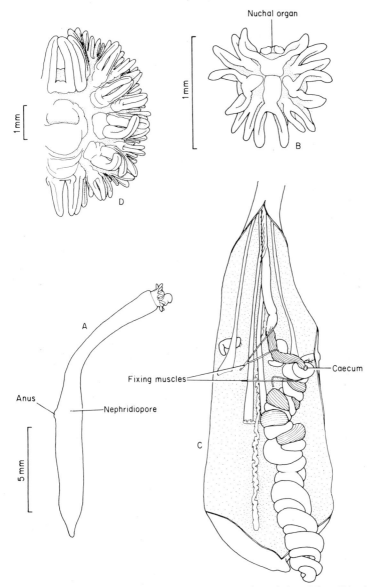

FIG. 3. *Golfingia margaritacea* A–C. Small specimens (13–18 mm in length from Wembury, Devon); A. Entire animal; B. Anterior view of oral disk; C. Internal structures of trunk; D. Large northern specimen from Trömso, Norway—anterior view of oral disk (after Théel, 1905).

Golfingia (Golfingia) vulgaris (de Blainville)

(Fig. 4A–D)

Sipunculus vulgaris de Blainville, 1827
Golfingia (Golfingia) vulgaris: Stephen and Edmonds, 1972: 110

A large species, up to 150–200 mm in length, with a cylindrical trunk (Fig 4A). Tentacular crown becomes more complex with increasing size; small individuals (10–15 mm long) have about 20 tentacles arranged in pairs in a single circle around mouth; with growth, further tentacles develop in radial longitudinal rows; when a body length of about 30 mm is reached, 50–60 are present (Fig. 4B) whilst in the largest specimens the tentacle count may total over 150. Simple, spine-like hooks are arranged irregularly over anterior introvert (Fig. 4C). Skin of introvert and trunk is finely wrinkled; at both ends of trunk it is rugose, has prominent papillae and is usually dark in colour. Nephridiopores are ventrolateral and just anterior of anus. Longitudinal muscle layer of body wall is continuous, not collected into bands.

Internally, four retractor muscles are present, the ventral pair being inserted in middle third of trunk and the dorsal pair well to posterior of anus (Fig. 4D). Intestine is tightly coiled in a double spiral supported by a spindle muscle attached anteriorly near anus, but not posteriorly. Rectal caecum conspicuous. Gut-fixing muscles fasten oesophagus and anterior coils of gut-spiral to body wall in the region between roots of dorsal and ventral retractors. Contractile vessel simple. Two nephridia; these hang freely in coelom.

Inhabits muddy sand/gravel, from lower shore to about 2000 m depth.

A widespread species in British waters. In the North Atlantic it ranges from Greenland and northern Norway to West Africa and the eastern Mediterranean. Also reported from scattered localities within the Indo-West Pacific region and from the Antarctic.

Type locality: Dieppe, France.

Notes. G. vulgaris is rather variable in its external appearance, hence many synonyms have been described and six or seven subspecies recognized. Abnormalities of the retractor system are less pronounced in this species than in *G. elongata* (p. 10); between 5 and 10% of individuals in some populations lack one of the dorsal retractors but the non-development of both dorsal retractors seems to be rare.

Frequently, *G. vulgaris* has large numbers of the entoproct *Loxosomella phascolosomata* (Vogt) attached to the trunk surface, particularly over the posterior region; the bivalve *Mysella bidentata* often inhabits its burrow. *G. vulgaris* spawns in the period June to September (Roscoff).

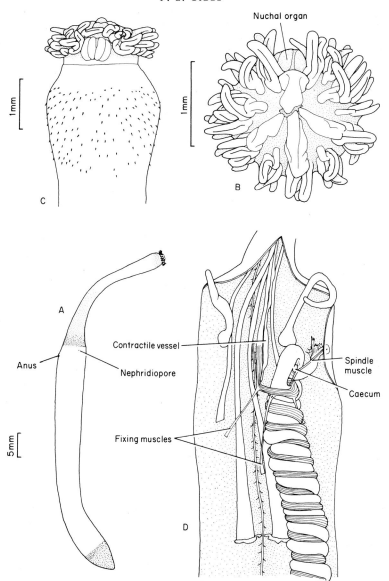

Fig. 4. *Golfingia vulgaris* A. Entire animal; B. Anterior view of oral disk of small specimen (30 mm in length) with 56 tentacles; C. Dorsal view of anterior introvert of same specimen; D. Internal structures of anterior trunk region.

Golfingia (Phascoloides) minuta (Keferstein)

(Fig. 5A–D)

Phascolosoma minutum Keferstein, 1862
Golfingia (*Phascoloides*) *minuta*: Stephen and Edmonds, 1972: 149

A small species, not exceeding 15 mm in length, with a cylindrical trunk (Fig. 5A). Oral disk carries a rudimentary tentacular crown consisting of a dorsal pair of short tentacles plus two to six rounded lobes or prominences which tend to be rather variable in form according to the degree of eversion (Fig. 5B). Simple, spine-like hooks may be present on anterior introvert; they are usually fairly numerous on juveniles (and regenerated specimens) but are few, or entirely absent, on adults, presumably through wear; when present, hooks are arranged irregularly (Fig. 5C). Skin of introvert and trunk is smooth but minute papillae may be discernible, especially over posterior trunk region. Nephridiopores are lateral and just posterior of anus on anterior trunk. Longitudinal muscle layer of body wall is continuous, not collected into bands.

Internally, two retractor muscles are present, inserted in middle third of trunk (Fig. 5D); often the two retractors are fused for most of their length. Intestine is tightly coiled in a double spiral supported by a spindle muscle attached anteriorly but not posteriorly. Rectal caecum usually conspicuous. Typically, one gut-fixing muscle fastens the oesophagus to body wall on left side.

Inhabits mud and sand, from mid-tide level to about 50 m depth (records from deeper waters are uncertain—see Notes below). Common in rock crevices, amongst *Sabellaria* tubes and similar niches.

A widespread species in European waters from Shetland and Sweden to Brittany (other records uncertain).

Type locality: St. Vaast la Hougue, France.

Notes. G. minuta is unusual, perhaps unique, amongst the Sipuncula in being hermaphroditic. Both sperm cells and oocytes are usually detectable in the coelom throughout the year. At Plymouth, it spawns from November to January. Sterility is caused by a parasitic copepod, *Akessonia occulta*, occurring singly in the coelom; this copepod, described from Swedish waters, has yet to be discovered in *Golfingia minuta* from British seas.

As a result of morphological variability, especially in the presence or absence of hooks and papillae, the taxonomy of G. minuta is very confused. G. improvisa (Théel), as described from Swedish waters, is here considered to be a junior synonym. The hermaphroditic form found in European seas appears to be distinct from a morphologically similar species found in deep water over a wide area of the North Atlantic; the latter is dioecious and probably referable to G. diaphanes (Gerould) (see Gibbs, 1975). The status of specimens ascribed to G. minuta from other regions remains uncertain.

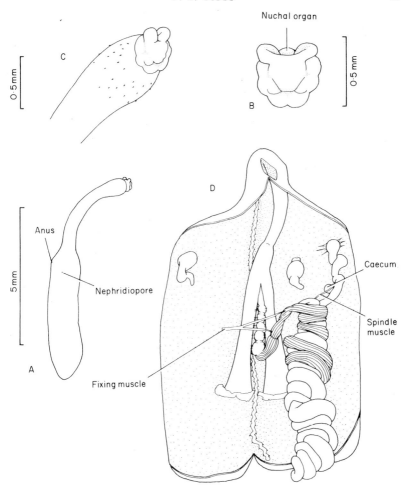

Fig. 5. *Golfingia minuta* A. Entire animal; B. Anterior view of oral disk; C. Lateral view of anterior introvert of young specimen with hooks; D. Internal structures (introvert withdrawn).

Golfingia (Phascoloides) rimicola Gibbs

(Fig. 6A–E)

Golfingia (Phascoloides) rimicola Gibbs, 1973: 74

A medium-sized species, up to 50–60 mm in length, with a slender, cylindrical trunk (Fig. 6A). Oral disk carries a single circle of paired tentacles surrounding mouth. Most adult specimens (>20 mm long) have 16 tentacles (Fig. 6B) but the number varies from eight (in juveniles) up to 20 in large specimens. Simple, spine-like hooks are arranged in well separated rings on anterior introvert (Fig. 6C); typically six to ten rings are present although there is considerable variation since the posterior rings are often incomplete or missing through wear. Skin of introvert and trunk is fairly smooth, lacking prominent papillae. Nephridiopores are ventrolateral, just anterior of anus on anterior trunk. Longitudinal muscle layer of body wall is continuous, not collected into bands.

Internally, two retractor muscles are present, inserted in anterior half of trunk (Fig. 6D). Intestine is tightly coiled in a double spiral supported by a spindle muscle attached anteriorly near anus, but not posteriorly. Rectal caecum usually conspicuous. Gut-fixing muscles not developed. Contractile vessel simple. Two nephridia; these hang freely in the coelom.

Inhabits mud and sand filling rock crevices on the lower shore where it is often found with *G. minuta* (p. 16).

G. rimicola is known only from the coasts of Devon and Cornwall.

Type locality: West Reef, Wembury, Devon.

Notes. G. rimicola may be confused with aberrant individuals of *G. elongata* having only two retractors instead of the normal four (p. 10). However, the two species are generally separable on the basis of the number of tentacles, about 92% of *G. rimicola* having 18 or less, 95% of *G. elongata* having 22 or more (Gibbs, 1973).

G. rimicola spawns in late October or November (Plymouth). Its burrow is occasionally inhabited by the bivalve *Epilepton clarkiae*.

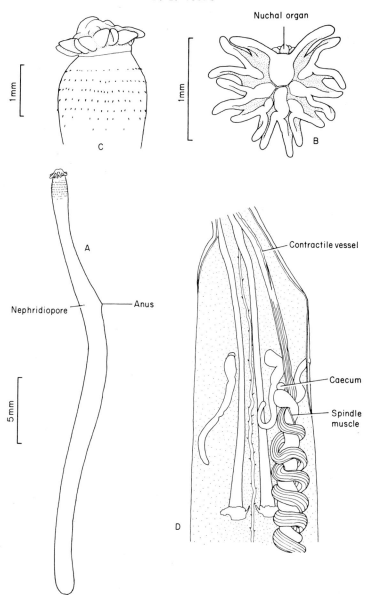

FIG. 6. *Golfingia rimicola* A. Entire animal; B. Anterior view of oral disk; C. Lateral view of anterior introvert; D. Internal structures of anterior trunk.

Golfingia (*Thysanocardia*) *procera* (Möbius)

(Fig. 7A–F)

Phascolosoma procerum Möbius, 1875
Golfingia (*Thysanocardia*) *procera*: Stephen and Edmonds, 1972: 129

A small- to medium-sized species, up to 50–60 mm in length, with a short tapering trunk and long, slender introvert which, when fully everted, is two to four times length of trunk (Fig. 7A). Oral disk carries fairly numerous tentacles surrounding mouth and arranged in radial/longitudinal rows; large specimens have 60–70 tentacles (Figs 7B, C). No hooks on introvert. Skin of introvert and trunk is finely wrinkled and has minute papillae; characteristically, the skin of the trunk is corrugated showing an irregular pattern of longitudinal zig-zag folds (Fig. 7D). Nephridiopores are ventrolateral and anterior of anus on anterior trunk. Longitudinal muscle layer of body wall is continuous, not collected into bands.

Internally, two retractor muscles are present; these are fused for most of their length and are inserted in posterior third of trunk (Fig. 7E). Intestine is tightly coiled in a double spiral supported by a spindle muscle attached anteriorly near anus, but not posteriorly. Rectal caecum present. One gut-fixing muscle fastens anterior coils of gut to dorsal body wall. Contractile vessel is conspicuous in that its distal half carries numerous branching villi (Figs 7E, F). Two nephridia these hang freely in coelom.

Inhabits muddy sand, in depths of 2–200 m.

In the north-east Atlantic, *G. procera* has a limited distribution extending from the Skagerrak and northern North Sea to the west coasts of Scotland and Ireland including the Irish and Celtic Seas. It appears to be absent from the southern North Sea and the English Channel. Elsewhere the species is recorded from off the east coast of North America between 34 and 40°N and off California (see Notes below).

Type locality: off Bass Rock, Firth of Forth, Scotland.

Notes. The distribution of the species outside of the north-east Atlantic region is uncertain. Cutler (1973) has suggested that *G. procera* and the closely related *G. semperi* (Selenka and de Man) are junior synonyms of *G. catherinae* (Grube) described from Brazil; such a combination gives the European form a world-wide distribution.

G. procera spawns in May and June (Milford Haven area). In the Kattegat population, individuals appear to act as temporary parasites of the polychaete *Aphrodita aculeata* L. (Thorson, 1957).

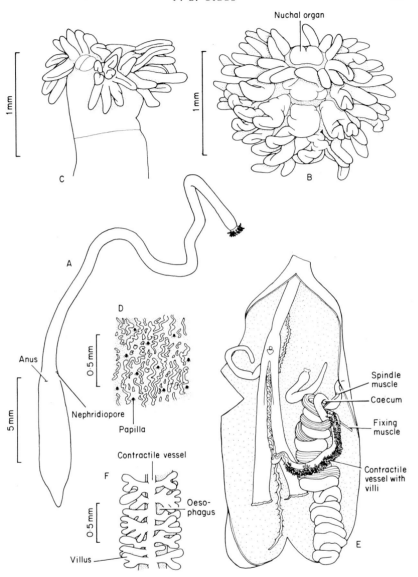

Fig. 7. *Golfingia procera* A. Entire animal; B. Anterior view of oral disk of specimen with about 65 tentacles; C. Dorso-lateral view of anterior introvert; D. Zig-zag figurations of trunk surface; E. Internal structures of trunk (introvert withdrawn); F. Part of contractile vessel near distal end showing branched villi.

Phascolion strombi (Montagu)
(Fig. 8A–E)

Sipunculus strombus Montagu, 1804
Phascolion strombi: Stephen and Edmonds, 1972: 187

A medium-sized species, up to about 50 mm in length, typically found living in mollusc shells (Fig. 8A). Its form is very variable, the shape and diameter depending on the shelter inhabited. Oral disk carries tentacles surrounding mouth; small specimens (<25 mm long) have a single circle of eight to 16 tentacles but with growth, further tentacles develop to form short, radial/longitudinal rows and large specimens have 40–50 tentacles (Fig. 8B). Simple, spine-like hooks are arranged irregularly on anterior introvert (Fig. 8C). Whole body surface is covered with papillae of varying sizes; over base of introvert and anterior trunk, the papillae are large and broadly conical in shape, whilst over the middle trunk region they are modified to form chitinous "holdfasts" (="adhesive" or "attachment" papillae); "holdfasts" are rather variable in appearance but typically are dark-coloured and ∧- or ∩-shaped (Fig. 8D). Single nephridiopore is posterior to prominent anus on anterior trunk. Longitudinal muscle layer of body wall continuous, not collected in bands.

Internally, two retractor muscles, of unequal size, are inserted at posterior end of trunk; dorsal retractor has a wide, undivided base whilst ventral retractor has two roots attached on either side of nerve cord (Fig. 8E). Intestine is arranged in a series of longitudinal loops fastened by a number of fixing muscles; its middle section is usually weakly coiled in a double spiral without a spindle muscle. Contractile vessel is simple and relatively large, often about same diameter of adjacent oesophagus. Rectal caecum present. Right nephridium only is developed; this is fastened to body wall for most of its length.

Found in mud and sand, in depths of 4–3800 m, usually inhabiting a wide variety of shelters, principally gastropod and scaphopod shells. In European seas, the shells of *Turritella*, *Aporrhais*, *Nassarius* and *Dentalium* are commonly utilized.

P. strombi is one of the commonest species in European waters and is widely distributed throughout the Atlantic Oceans. Also reported off southern Chile and in the Red Sea.

Type locality: south coast of Devonshire.

Notes. The external features of *P. strombi* are subject to great variations which, to some degree, appear to be influenced by environmental conditions. Many forms, now considered to be conspecific, have been described as separate species or as distinct varieties (see Stephen and Edmonds, 1972). A closely related species, *P. tuberculosum* Théel, is found in Scandinavian waters but has yet to be recorded from around the British Isles; it is chiefly distinguished by its strongly curved hooks and lack of holdfast papillae. *Phascolosoma intermedium* Southern (1913b) is almost certainly based on juveniles of *P. strombi* (Gibbs, 1977).

Shells inhabited by *P. strombi* are readily recognizable since the aperture is partially blocked by a plug of cemented sediment particles with a central hole through which the introvert is protruded. The shelter is often inhabited by a variety of other animals, notably the polychaete *Syllis* (*Langerhansia*) *cornuta* the parasitic gastropod *Menestho diaphana* (see Kristensen, 1970) and, more rarely, the bivalve *Tellimya phascolionis*. Entoprocts of the genus *Loxosomella* are commonly attached to the trunk surface, especially the holdfasts. In Swedish waters, *P. strombi* has a prolonged spawning season perhaps extending throughout the year (Åkesson, 1958).

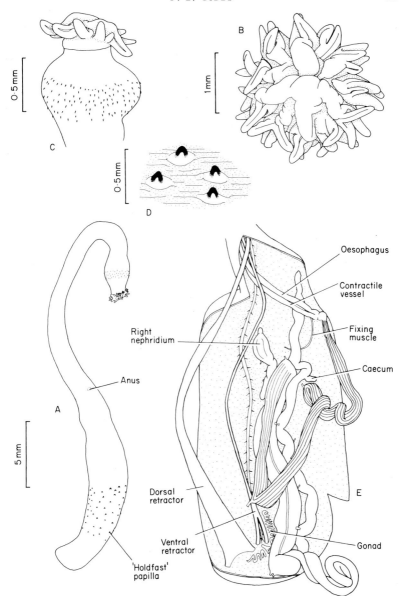

Fig. 8. *Phascolion strombi* A. Entire animal (removed from gastropod shell before fixation); B. Anterior view of oral disk of specimen with about 36 tentacles; C. Lateral view of anterior introvert of young specimen (15 mm in length); D. Chitinous 'holdfast' papillae on trunk; E. Internal structures of trunk.

Onchnesoma squamatum (Koren and Danielssen)

(Fig. 9A–D)

Phascolosoma squamatum Koren and Danielssen, 1876
Onchnesoma squamatum: Stephen and Edmonds, 1972: 163

A small species, up to 30 mm in length, with a short, pear-shaped trunk typically 2–7 mm in length and a long, slender introvert about three times length of trunk (Fig. 9A). Tentacular crown is simple consisting of about eight small (primary) tentacles arranged in a single circle around mouth (Fig. 9B). Introvert lacks hooks but has minute papillae. Whole surface of trunk is covered with backwardly directed, scale-like protrusions of irregular shape and varying size (Fig. 9C). Single nephridiopore is ventrolateral on anterior trunk whilst anus is removed to anterior half of introvert. Longitudinal muscle layer of body wall is continuous, not collected into bands.

Internally, one retractor muscle is present, inserted at the posterior end of trunk; two roots are usually distinguishable (Fig. 9D). The oesophagus is attached to the ventral surface of the retractor muscle for most of the latter's length; proximal part of the intestine makes several longitudinal loops, held by fixing muscles, before coiling in a double spiral supported by a thin spindle muscle; rectum is long extending into introvert when everted to the anteriorly placed anus. Rectal caecum present. Contractile vessel not discernible. Right nephridium only is developed; this is attached to the body wall for most of its length.

Inhabits mud and sands in depths of 180–1000 m.

The species is widely distributed along the edge of the continental shelf from northern Norway (Lofoten Is.) and Iceland to Portugal. In the western Atlantic region it occurs between latitudes 34–24°N. Also reported from eastern Mediterranean in shallow depths (18–56 m).

Type locality: Hardangerfjord, Norway.

Notes. Little is known of the biology of this distinctive species. It is often taken in large numbers, along with the closely related *O. steenstrupi* (p. 26).

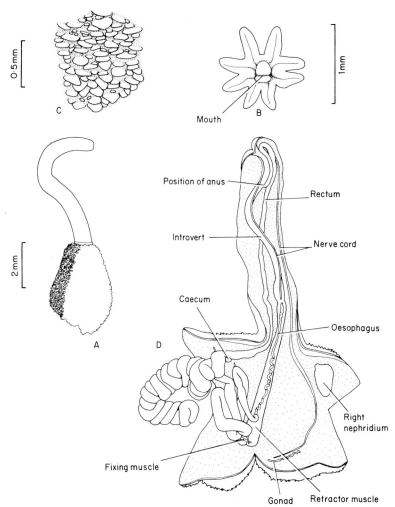

FIG. 9. *Onchnesoma squamatum* A. Entire animal (introvert not fully extended); B. Anterior view of oral disk (after Théel, 1905); C. Surface of anterior trunk showing characteristic protrusions of skin; D. Internal structures.

Onchnesoma steenstrupi Koren and Danielssen

(Fig. 10A–D)

Onchnesoma steenstrupi Koren and Danielssen, 1876; Stephen and Edmonds, 1972: 163.

A small species, up to 40 mm in length, with a short, pear-shaped trunk typically 1–5 mm in length, and a long, slender introvert (Fig. 10A). Tentacular crown not developed as such since oral disk is without tentacles or lobes (Fig. 10B). Introvert lacks hooks but has minute papillae; trunk is transversely wrinkled and covered with flat to almost spherical "scales" between which are small papillae; over posterior trunk the scales are aligned and fused to form longitudinal ridges. Single nephridiopore is ventrolateral on anterior trunk whilst anus is removed to anterior introvert, close to the oral disk (Fig. 10C). Longitudinal muscle layer of body wall is continuous, not collected in bands.

Internally, one retractor muscle is present, inserted at posterior end of trunk (Fig. 10D). Oesophagus not fastened to retractor muscle as in *O. squamatum* (p. 24). Middle section of intestine is coiled in a double spiral (? without spindle muscle) whilst proximal and distal sections, and also oesophagus and rectum, are gathered into loose loops when the introvert is withdrawn; when everted, these sections traverse the length of introvert, the anus being close to the mouth. Rectal caecum present. Contractile vessel not apparent. Right nephridium only is developed; this is fastened to body wall for most of its length.

Inhabits mud and sand in depths of 25–900 m.

The species is widely distributed in the North Atlantic; in the north eastern region it ranges from northern Norway to the eastern Mediterranean and in the western region from 40–30°N.

Type locality: Moldefjord, near Kristiansund, Norway.

Notes. The biology of *O. steenstrupi* is poorly known. In deep water it occurs with the closely related *O. squamatum*.

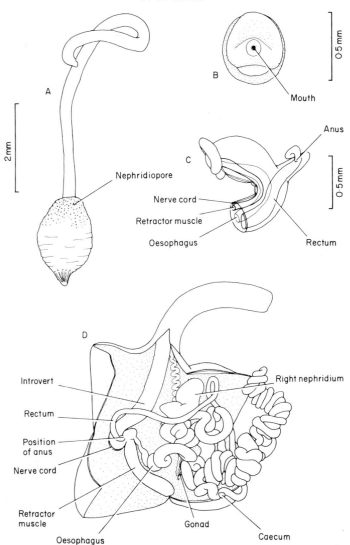

FIG. 10. *Onchnesoma steenstrupi* A. Entire animal (introvert not fully extended); Anterior view of oral disk; C. Lateral view of anterior introvert showing position of anus; D. Internal structures. (B & C after Théel, 1905).

28   Family PHASCOLOSOMATIDAE

Oral disk with tentacles arranged in a crescent enclosing nuchal organ. Coelomic canals or sacs not present in body wall. Longitudinal muscle layer usually collected into bands. Trunk without shields. Two nephridia. One species is found in British waters.

Phascolosoma granulatum Leuckart
(Fig. 11A–E)

Phascolosoma granulatum Leuckart, 1828; Stephen and Edmonds, 1972: 306

A medium- to large-sized species, up to 100 mm in length with a stout tapering trunk (Fig. 11A). Oral disk has a swollen rim (cephalic collar) and carries a horseshoe-shaped crescent of tentacles situated dorsally to mouth and enclosing a large nuchal organ (Fig. 11B); tentacles number from 12–16 in small specimens up to 30–60 in medium to large individuals. On anterior introvert there are numerous hooks, each with a curved tip and a wide triangular base (Fig. 11C) arranged in rings; the number of rings varies from 10–20 to 50–60 depending on size of specimen and often posterior rings are incomplete through wear. Whole body surface has dome-shaped papillae of varying size, each capped with a dark ring around a central clear spot; the largest papillae occur over base of introvert and over posterior trunk where they are conspicuous in being darker coloured (Fig. 11D). Nephridiopores are ventrolateral and just anterior of anus on anterior trunk. Longitudinal muscle layer of body wall is collected into 20–30 bands joined by numerous anastomoses (Fig. 11E); in small specimens these bands may not be readily apparent.

Internally, four retractor muscles are present, each inserted by several roots onto longitudinal muscle bands in the middle third of trunk (Fig. 11E); ventral retractors are thicker and have wider bases than dorsal retractors. Intestine is loosely coiled into a double spiral supported by a strong spindle muscle attached anteriorly near anus and also at posterior end of trunk. Rectal caecum usually small. Contractile vessel simple. A fixing muscle fastens oesophagus and rectum to body wall between the roots of dorsal retractors. Two nephridia; these are fastened to body wall over their anterior halves.

Inhabits muddy sand/gravel, lower shore to 90 m depth. Intertidally it occurs under stones, in crevices and amongst *Lithothamnion*.

P. granulatum has a restricted distribution in British seas in that it is common along the west coast of Ireland but elsewhere only a few specimens have been recorded, originating from Shetland, Orkney and the Hebrides. The species ranges from the coastal waters of Norway (one specimen) and France to the Cape Verde Is. and throughout the Mediterranean. It is also reported from scattered localities in the Indo-West Pacific region.

Type locality: Sète (Cette), southern France.

Notes. P. granulatum is rather variable in its external appearance and thus has many synonyms. In larger specimens the papillae are normally darker and more conspicuous than in small specimens; in the latter, the introvert often has dark transverse bands particularly over the dorsal surface. Spawning appears to occur in August or September (Donegal). The bivalve *Epilepton clarkiae* occurs in its burrows.

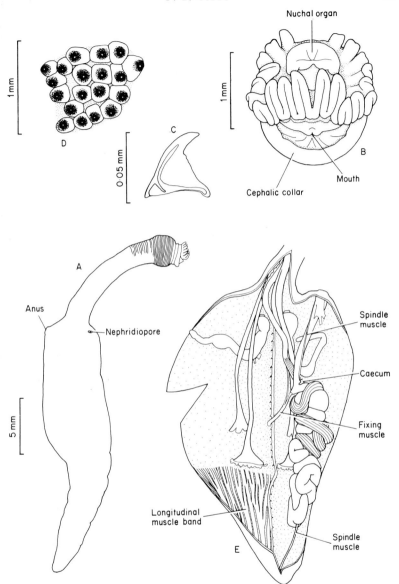

FIG. 11. *Phascolosoma granulatum* A. Entire animal; B. Anterior view of oral disk; C. Side view of hook; D. Skin papillae at base of introvert; E. Internal structures of trunk (longitudinal muscle bands shown for left posterior region only).

Family ASPIDOSIPHONIDAE

Oral disk with short tentacles arranged in a crescent enclosing nuchal organ
Coelomic canals or sacs not present in body wall. Longitudinal muscle layer
either continuous or in bands. Anterior end of trunk chitinized to form anal
shield; posterior end usually likewise modified to form caudal shield. Two
nephridia. A single species occurs in British waters.

Aspidosiphon muelleri Diesing
(Fig. 12A–D)
Aspidosiphon muelleri Diesing, 1851; Stephen and Edmonds, 1972: 231

A medium- to large-sized species, up to 80 mm in length with dark chitinous
shields at both ends of a cylindrical trunk (Fig. 12A). Oral disk has a swollen
rim (cephalic collar) and carries 10–12 tentacles arranged in a crescent enclosing
nuchal organ; tentacles are short and united for much of their length (Fig.
12B). Introvert arises from ventral margin of anal shield and is densely covered
with small hooks; anteriorly the hooks are thin, triangular in outline and
arranged in numerous rings (Fig. 12C) becoming more conical, spine-like
and arranged irregularly over posterior half or two-thirds of introvert. Anal
(anterior) and caudal (posterior) shields of trunk are heavily chitinized areas
composed of closely packed, rounded or polygonal bodies; anal shield is radially
furrowed over its dorsal half, with raised, wart-like papillae ventrally; caudal
shield has radial furrows and a peripheral zone of prominent papillae. Nephridiopores are ventrolateral and at about level of anus on anterior trunk. Longitudinal muscle layer of body wall is continuous except in the anal shield region
where it forms tendinous bands.

Internally, one retractor muscle is present, inserted by two roots on ventral
margin of caudal shield (Fig. 12D). Intestine is tightly coiled in a double spiral
supported by a spindle muscle attached anteriorly near anus and posteriorly on
the dorsal part of caudal shield. A long fixing muscle fastens anterior coils of
gut to body wall, running through the cleft in retractor muscle. Rectal caecum
small. Contractile vessel not discernible. Two nephridia; each is attached
throughout its length.

This sipunculan is found in a wide variety of habitats; it commonly occupies
gastropod and scaphopod shells and serpulid tubes but also lives in crevices,
amongst coralline algae and in deepwater corals (*Lophohelia*), from lower shore
to a depth of about 1000 m.

Widely distributed in the eastern Atlantic from Shetland and Norway to West
Africa and eastern Mediterranean. Also reported from scattered localities in the
Indo-West Pacific region.

Type locality: off Palermo (Panormum), Sicily.

Notes. With age, *A. muelleri* becomes darker in colour and large specimens
are frequently dark brown or even black. In Atlantic specimens all hooks have
one point but the Mediterranean form is characterized by the anterior-most
hooks having two points.

When the introvert is withdrawn the anal shield effectively serves as an oper-
culum blocking the shell aperture or burrow opening. This species does not plug
the shell aperture with sediment as is done by *Phascolion strombi*. The polychaete
Syllis (*Langerhansia*) *cornuta* is a frequent co-habitant in shells occupied by *A.
muelleri*.

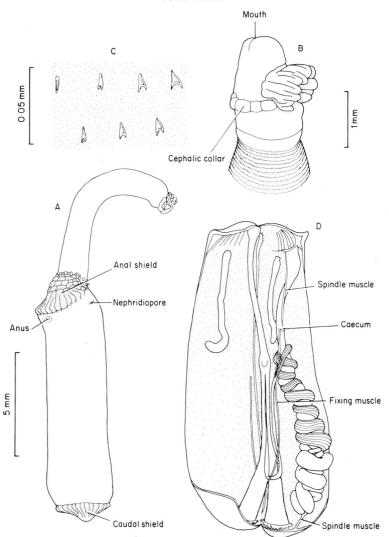

FIG. 12. *Aspidosiphon muelleri* A. Entire animal; B. Dorso-lateral view of anterior introvert (slightly over-everted); C. Hook circles on anterior introvert; D. Internal structures (introvert withdrawn).

Family SIPUNCULIDAE

Oral disk carries a large tentacular lobe surrounding mouth; margin of lobe may be sinuous, scalloped or digitate. Body wall of trunk region with coelomic canals or sacs. Longitudinal muscle layer collected into bands. Trunk without shields. Two nephridia. One species occurs in British waters.

Sipunculus nudus Linnaeus
(Fig. 13A–F)

Sipunculus nudus Linnaeus, 1766; Stephen and Edmonds, 1972: 32

A large species, up to about 350 mm in length, with a cylindrical trunk (Fig 13A). Oral disk carries a large tentacular lobe which is highly folded, its margin sinuous or scalloped but not markedly incised to form tentacles (Fig. 13B, C). Introvert is without hooks but is covered with large flattened papillae which are triangular in shape and backwardly-pointing (Fig. 13C, D); surface of trunk has characteristic pattern of rectangles defined by grooves formed by the underlying circular and longitudinal muscle bands (Fig. 13A, E) but this pattern may not be discernible in small specimens (<35 mm long). Nephridiopores ventrolateral and anterior to anus on anterior trunk. Body wall thick with circular, diagonal and longitudinal muscle layers collected into bands (Fig. 13F) and containing longitudinal canals which communicate with coelom at intervals. Longitudinal muscle bands number 28–34 in anterior trunk region, fewer in posterior trunk.

Internally, four retractor muscles are present, all inserted in the anterior trunk at about level of anus, their bases each extending across four to eight muscle bands (Fig. 13G). Intestine is loosely coiled in a double spiral fastened to body wall by numerous fixing muscles and supported by a spindle muscle attached anteriorly near anus, but not posteriorly; intestine also has a characteristic minor coil (the so-called "post-oesophageal" or "*Sipunculus*-loop") between oesophagus and main spiral. Rectal caecum variable in size but usually small. Two simple contractile vessels on oesophagus, one dorsal and one ventral. Two nephridia; these are attached for about one quarter of their length. A pair of "racemose" glands is held in thin mesenteries, one on either side of rectum.

Inhabits sand, from lower shore to about 700 m depth (the specimen from 2310 m, identified by Selenka (1885) as this species, is in fact, *S. norvegicus*).

In European waters *S. nudus* is found from the west coast of Ireland and southern North Sea to the eastern Mediterranean. It has a worldwide distribution, mainly in temperate and tropical seas.

Type locality: unknown.

Notes. A closely related species, *S. norvegicus* Danielssen, occurs in the deeper waters of the north eastern Atlantic but since it has not been recorded from the shelf zone it is not described here. It is chiefly distinguished from *S. nudus* in having fewer longitudinal muscle bands (22–24).

FIG. 13. *Sipunculus nudus*. A. Entire animal; B. Anterior view of oral disk; C. Lateral view of anterior introvert; D. Scale-like papillae on introvert; E. Rectangular figuration of trunk skin; F. Internal surface of trunk wall showing arrangement of muscle bands; G. Internal structures of anterior trunk (muscle bands of trunk wall not illustrated).

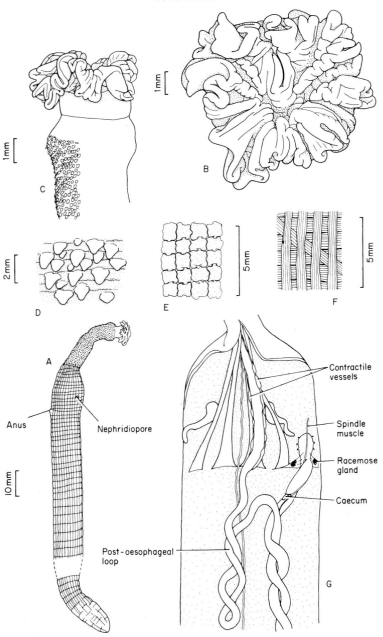

A

Anus

Nephridiopore

10mm

Post-oesophageal
loop

B

C

D

E

F

G

Contractile
vessels

Spindle
muscle

Racemose
gland

Caecum

34 References

ÅKESSON, B. 1958. A study of the nervous system of the Sipunculoideae, with some remarks on the development of the two species *Phascolion strombi* Montagu and *Golfingia minuta* Keferstein. *Unders. över Öresund*, **38**, 249pp. Lund: Gleerup.

CUÉNOT, L. 1922. Sipunculiens, Échiuriens, Priapuliens. *Faune Fr.* **4**, 1–29.

CUTLER, E. B. 1973. Sipuncula of the western North Atlantic. *Bull. Am. Mus. nat Hist.* **152**, 103–204.

CUTLER, E. B. and MURINA, V. V. 1977. On the sipunculan genus *Golfingia* Lankester, 1885. *Zool. J. Linn. Soc.* **60**, 173–187.

FISCHER, W. 1925. Echiuridae, Sipunculidae, Priapulidae. *Tierwelt N.-u. Ostee*, **1(VId)** 1–55.

FISHER, W. K. 1950. The sipunculid genus *Phascolosoma*. *Ann. Mag. nat. Hist.* **(12) 3**, 547–552.

FORSTER, G. R. 1953. A new dredge for collecting burrowing animals. *J. mar. biol Ass. U.K.* **32**, 193–198.

GEROULD, J. H. 1906. The development of *Phascolosoma*. (Studies on the Embryology of the Sipunculidae II.) *Zool. Jb.* (Anat.) **23**, 77–162.

GIBBS, P. E. 1973. On the genus *Golfingia* (Sipuncula) in the Plymouth area with a description of a new species. *J. mar. biol. Ass. U.K.* **53**, 73–86.

GIBBS, P. E. 1974. *Golfingia margaritacea* (Sipuncula) in British waters. *J. mar. biol. Ass. U.K.* **54**, 871–877.

GIBBS, P. E. 1975. Gametogenesis and spawning in a hermaphroditic population of *Golfingia minuta* (Sipuncula). *J. mar. biol. Ass. U.K.* **55**, 69–82.

GIBBS, P. E. 1977. On the status of *Golfingia intermedia* (Sipuncula). *J. mar. biol. Ass. U.K.* **57**, 109–112.

HYMAN, L. H. 1959. *The Invertebrates. V. Smaller coelomate groups*. 783pp. New York: McGraw-Hill.

KRISTENSEN, J. H. 1970. Fauna associated with the sipunculid *Phascolion strombi* (Montagu), especially the parasitic gastropod *Menestho diaphana* (Jeffreys). *Ophelia* **7**, 257–276.

MURINA, V. V. 1975. Ways of evolution and phylogeny of the Phylum Sipuncula. *Zool. Zh.* **54**, 1747–1758 [In Russian with English Summary].

RICE, M. E. 1975. Sipuncula. In *Reproduction of Marine Invertebrates*. (Ed. A. C. Giese and J. S. Pearse) Vol. II. *Entoprocts and Lesser Coelomates*, pp. 67–127. New York and London: Academic Press.

SELENKA, E. 1885. Report on the Gephyrea collected by H.M.S. Challenger during the years 1873–76. *Rep. scient. Results Voy. "Challenger"* (Zool.) **13 (36)**, 1–24.

SOUTHERN, R. 1913a. Gephyrea of the coasts of Ireland. *Scient. Invest. Fish. Brch Ire.* 1912, No. III, 1–46.

SOUTHERN, R. 1913b. Survey of Clare Island. Part 49. Gephyrea. *Proc. R. Ir. Acad.* **31(49)**, 1–6.

STEPHEN, A. C. 1964. A revision of the classification of the Phylum Sipuncula. *Ann. Mag. nat. Hist.* **(13) 7**, 457–462.

STEPHEN, A. C. and EDMONDS, S. J. 1972. *The Phyla Sipuncula and Echiura*. 528pp. London: British Museum (Natural History).

TÉTRY, A. 1959. Classe des Sipunculiens. In *Traité de Zoologie* (Ed. P.-P. Grassé). **V.** pp. 785–854. Paris: Masson et Cie.

THÉEL, H. 1905. Northern and Arctic invertebrates in the Collection of the Swedish State Museum (Riksmuseum). I. Sipunculids. *K. svenska VetenskAkad. Handl.* **39(1)**, 1–130.

THORSON, G. 1957. Parasitism in the Sipunculid, *Golfingia procerum* (Moebius). *J. Fac. Sci. Hokkaido Univ.* (Zool.) **(VI), 13**, 128–132.

WESENBERG-LUND, E. 1930. Priapulidae and Sipunculidae. *Dan. Ingolf-Exped.* **IV(7)**, 1–44.

Index of Species

The correct generic and species names are shown in italics, synonyms in roman.
The page citations in roman refer to the text; those in italics are to illustrations.

Notes

Notes